Energy Harvesters

The Future of Sustainable Power

Table of Contents

Chapter 1. Introduction

In this Special Report, we invite you to explore the fascinating world of Energy Harvesters – the next frontier in power generation and sustainability. Written in accessible language and geared toward both experts and generalists, this report dives deep into this cutting-edge technology that seeks to harness the omnipresent energy around us and convert it into usable power. Will these devices truly herald a revolution in sustainable power? How close are we to seeing them integrated into our daily lives? Our detailed analysis will ensure that you grasp the science at the heart of this exciting subject, while also understanding the very real-world significance it holds. Invest in this inclusive report today and get a head start on the future of sustainable power. Your journey through the bustling streets of energy evolution is just a purchase away!

Chapter 2. The Basics of Energy Harvesting

Energy harvesting, often referred to as power harvesting or energy scavenging, comprises techniques that capture and store small amounts of energy from the surrounding environment. This energy, often derived from sources such as waste heat, light, vibration, and radio waves, is then converted into electrical power which can be used by low-power devices such as sensors and remote wireless networks.

2.1. From Scavenging to Harvesting

While the term 'energy scavenging' might imply scrounging around for leftovers, it actually is a fundamental concept in the field of energy harvesting. Since many sources emitting energy tend to do so in a continuous manner, there is a constant stream of 'leftover' energy that is discarded on a regular basis. Energy scavenging, thus, seeks to capitalize on these sources and convert them into usable power, a task that requires innovative techniques and methods.

One such technique involves scavenging vibrational energy. In any environment, there are typically a number of vibrating components that generate energy - whether in the home, the office, or any other setting, mechanical vibrations are a constant. Energy harvesters can capture these vibrational forces and transform them into electricity.

In a similar vein, thermal energy harvesting makes use of temperature differentials to produce power. Given that temperature variations are ubiquitous, thermal energy provides a valuable resource for energy harvesting technologies.

2.2. The Science Behind Energy Harvesting

To understand how energy can be 'harvested', it's essential to delve into the physics behind the process. The energy harvesting system consists mainly of three parts: the transducer, power management, and the storage device.

A transducer is a device that transforms one form of energy into another. For instance, a solar photovoltaic (PV) cell converts sunlight (or light energy) into electrical energy. Similarly, a piezoelectric material can generate voltage when it is subjected to mechanical stress or vibrations, and a thermoelectric generator can convert waste heat into electricity.

Power management circuits serve to extract, condition, and manage the power harvested by the transducer. More precisely, they aim to optimize the harvested power, match the electrical properties of the transducer and the electrical application, and provide a constant power output despite the fluctuations of the harvested energy.

The storage device – typically a battery or a capacitor – stores the generated power. Given the intermittent nature of many energy sources, it's essential to have a method for storing the power that's produced and ensuring it can be used when required.

2.3. Materials in Energy Harvesting

Advancements in materials science have played a critical role in energy harvesting technologies. From piezoelectric materials that produce electricity in response to mechanical deformation to thermoelectric materials that can convert heat directly into electricity or vice versa, these innovations cover a broad array of applications and make energy harvesting increasingly accessible even for small-scale applications.

Specialized materials, called photovoltaic materials, directly convert light, generally sunlight, into electricity. The most common example of these materials is the silicon used in most solar panels today. Another example would be materials such as organic polymers or perovskites that are giving rise to new possibilities for flexible, lightweight, and inexpensive solar harvesters.

2.4. Harnessing the Power of the Environment

A variety of energy sources are available in the environment from which energy can be scavenged, such as light (solar), thermal (temperature gradients), kinetic (motion), and wireless radiofrequency (RF) signals.

Solar energy, one of the most abundant energy sources on Earth, can be harvested during day time with the help of solar cells. Although this source provides a massive amount of energy, its main disadvantage lies in its availability, which is dependent on weather conditions and day-night cycles.

Harvesting energy from radio frequency signals is a relatively new concept. With our world now teeming with electromagnetic signals, RF scavengers are uniquely poised to tap into this abundant energy source.

2.5. The Role of Energy Harvesting in Tomorrow's World

While the concept and practice of energy harvesting are not new, ongoing advancements in technology and materials science have enabled its integration into multiple devices and services. From powering wearable devices to generating power for remote sensors in hostile environments, energy harvesting technologies are transforming our everyday lives.

Furthermore, as the world becomes more connected and the Internet of Things (IoT) continues to expand, the demand for energy-efficient, low-power devices will surely rise. Here's where energy harvesters, with their ability to prolong the battery life of wireless devices, or even replace batteries in some cases, can shine.

Not only does energy harvesting offer a promising solution to powering our increasing number of devices, but it also signals a significant leap forward in sustainable energy generation. By harnessing 'waste' energy and repurposing it, we are taking a crucial step towards a more sustainable, efficient energy future.

Energy harvesting enables a world where power is available anywhere, anytime and to anyone. By capturing and reusing energy that would otherwise be wasted, it offers an invaluable opportunity to contribute to global sustainability while meeting our ever-growing energy demands. In this new frontier, efficient, sustainable energy is within our grasp.

Tangible advances in energy harvesting technologies are a testament to the incredible potential locked within our environment. As we innovate and evolve, our ability to utilize this potential will only grow stronger, paving the way for a new era of power generation and sustainability.

Chapter 3. History and Evolution of Energy Harvesters

To comprehend the significance of energy harvesters, it's essential to understand their history and evolution. The field of energy harvesting is not entirely new—humans have always sought to harness energy from the environment. However, the complexity, the tools, and the principle applications have evolved significantly over the centuries.

The earliest forms of energy harvesting can be traced back to the invention of fire and primitive tools. Fire provided the first sustained source of energy humans could use for warmth, light, and cooking. This marked our first true interaction with energy harvesting, as the hominids captured energy stored within wood and transformed it into heat and light, a process we now understand as combustion.

Later on, as civilization progressed, we witnessed the formation of larger scale energy harvesting means. Windmills were first invented around 2000 B.C., and used wind to grind grain or pump water. Water wheels employed the power of running water to grind flour. These methods offered larger, more efficient scales of energy harvesting, and showed us humans' increasing capability to manipulate and transform nature's energy sources.

3.1. The Industrial Revolution and the Age of Steam

While the first instances of energy harvesting were mostly mechanical, the birth of the Industrial Revolution brought with it a radical shift in the mode of energy harvesting. The transition from wood and other biomass to coal marked a significant departure from earlier methods. Coal enabled us to power steam engines, heat homes, and later produce electricity, and thus the age of steam began.

This momentous period in history also saw the rise of fossil fuels as powerful energy sources. As time passed, humans developed the capability to exploit oil and natural gas, transforming these raw materials into electricity and fuel for automobiles and industry. This represents the second phase in the history of energy harvesting, where the emphasis shifted from mechanical systems to electrical and combustion processes.

3.2. The Birth of Microenergy Harvesting

Fast forward to the 20th century, where technology advanced rapidly and opened up fresh avenues for smaller scale energy harvesting. This period marked the advent of what we consider modern energy harvesting—the ability to harness energy from ambient sources such as solar, wind, thermal and kinetic energy to power micro and nano devices. This has come to be known as microenergy harvesting.

After the invention of the transistor in 1947, the need for smaller, portable energy sources became more pronounced. The invention of the solar cell in the 1950s was a game-changer in the field of energy harvesting. No longer was it necessary to have a large scale system (like a windmill or a steam engine); instead, a small device could use sunlight to generate electricity, marking the beginning of a new era where every device had the potential to generate its own power.

3.3. The Advent of Piezoelectric Energy Harvesting

One of the most significant advancements in the field of energy harvesting came with the development of piezoelectric materials in the mid-20th century. These materials generate electricity when subjected to mechanical stress, thus enabling the conversion of kinetic energy into electrical energy.

This was a ground-breaking development in the field, opening doors to countless new possibilities. From powering the quartz watches to supplying micro power to IoT devices, the advent of piezoelectric energy harvesting represented a major evolutionary step in the history of energy harvesting.

3.4. The Present and the Future

Today, we are on the cusp of another energy revolution. With the increasing importance of sustainability and CO_2 emission reduction, the focus has once again shifted, this time towards renewable energy sources. Energy harvesting devices, due to their innate ability to convert ambient energy into usable power, hold the key to this transition.

We are now witnessing the integration of energy harvesting devices in a wide range of applications, from self-powered sensors to smart fabrics, implantable medical devices, and even skyscrapers. The future of energy harvesting looks promising, and the field is undergoing continuous evolution as it embraces micro-energy harvesting principles.

The exciting blend of nanotechnology, materials science and electrical engineering is now pushing the envelope of what's possible in energy harvesting. Today, the versatility and the miniaturization of these devices are the primary focus, with the objective to make every device energy self-sufficient. Energy harvesting is moving from merely supporting our needs to being intricately woven into the very fabric of our everyday lives. Hence, it's safe to say that the history and evolution of energy harvesters is far from over.

While the past of energy harvesting gives us valuable insights, its future is ripe with possibilities waiting to be discovered. As technologies continue to evolve, the day isn't far when energy harvesters become an ubiquitous part of our daily lives, changing the way we perceive power generation and consumption. This is the next frontier in our pursuit of sustainability, and in our journey towards a cleaner, greener planet.

Chapter 4. Types of Energy Harvesters and How They Work

Energy harvesting is an ingenious concept that profoundly changes our approach to power generation, converting ambient energy sources into usable electrical power. Let's take a closer look at the major types of energy harvesters and unpack how each one works.

4.1. Mechanical Energy Harvesters

Mechanical energy harvesters convert kinetic energy from physical movements into electrical power. These devices come in a range of forms depending on their application.

4.1.1. Piezoelectric Harvesters

Piezoelectric harvesters operate on the principle of piezoelectricity. When subjected to stress, certain types of materials generate an electric charge. In these harvesters, mechanical movement or vibration provides the stress to generate electricity.

Having no moving parts, piezoelectric harvesters are reliable and efficient. They are ideal for environments with high-frequency vibrations. A common application is within machinery to capture wasted kinetic and vibrational energy. However, their output is limited, offering only enough power for low-energy applications such as sensors.

4.1.2. Electromagnetic Induction Harvesters

These harvesters operate under the concept of electromagnetic induction, where a change in magnetic field induces an electric field. Simply put, as a conductive material moves within a magnetic field, it generates an electrical current.

Due to their high power output and robust nature, they're often found in larger-scale applications, such as wind turbines and wave energy converters. However, they contain moving parts, which may require maintenance over time.

4.2. Thermal Energy Harvesters

Without efficient energy management, wasted heat presents not only an environmental issue but also a missed opportunity for energy generation. Thermal energy harvesters capture and utilize this wasted heat.

4.2.1. Thermoelectric Harvesters

Thermoelectric harvesters exploit the Seebeck effect, where a temperature difference across a conductive material induces a voltage. Utilizing this principle, these harvesters convert waste heat into electricity, often supplementing power for machinery or powering low-energy devices.

They have no moving parts, making them practical and reliable, but their efficiency depends greatly on the temperature differential. Applications include industrial processes with high waste heat and body heat capture in wearable tech.

4.3. Solar Energy Harvesters

Solar energy harvesters, more commonly known as solar panels, need little introduction. These devices convert radiant solar energy into electric power using the photovoltaic effect.

Photovoltaic cells, integral to solar panel design, consist of layers of semiconductor material. On solar irradiation, they release electrons, creating an electrical current. Applications are vast, from small-scale electronics to rooftop installations and solar farms.

4.4. Radiowave Energy Harvesters

A relatively new and exciting branch of energy harvesting involves capturing energy from radiowaves. TV, radio, Wi-Fi signals, and even microwave transmissions we bath in daily carry energy that's predominantly wasted.

Radiowave energy harvesters capture some of this omnipresent ambient energy. The principle is akin to a radio receiver, where tuned circuits resonate at specific frequencies, capturing and rectifying this energy into DC power.

While their power output is currently low, innovations promise to improve their usefulness. Ideal applications are places with strong signal densities like cities.

4.5. Biochemical Energy Harvesters

As the name suggests, these devices harness energy from biochemical reactions.

4.5.1. Enzymatic Biofuel Cells

Enzymatic biofuel cells generate electricity through the enzymatic oxidation of biofuels. These devices have exceptional promise for wearables, drawing power from bodily fluids, including sweat.

However, these energy harvesters remain in the development stage, with reliability and lifetime being common challenges.

4.6. Hybrid Energy Harvesters

Harnessing energy from multiple sources, hybrid energy harvesters are designed to maximize power output and efficiency. They often combine techniques such as solar and piezoelectric energy harvesting, ensuring a more reliable output.

While they're relatively complex and may not be suitable for all applications, their ability to capture energy in a variety of conditions makes them highly potent.

Understanding this multitude of energy harvesting techniques is crucial for appreciating the breadth, depth, and potential of this technology. It offers us a greener, more sustainable future, whether by powering smart IoT devices, contributing to the grid, or enabling self-sustaining, off-grid systems. Through continuous innovation, this field will undoubtedly play a key role in shaping the sustainable energy landscape. As we venture deeper into the age of low-power electronics and clean energy policies, these energy harvesters hold the beacon high for a new epoch where power is sourced not from exhaustible resources, but from the limitless energy around us.

Chapter 5. Harvesting Solar, Wind and Kinetic Energy: Comparative Analysis

Comparative analysis provides a clear perspective on harnessing energy from the most predominant sources: solar, wind, and kinetic. These renewable and omnipresent forms of energy, when tapped and utilized efficiently, will illuminate the path to an eco-friendly and sustainable future.

5.1. Solar Energy Harvesting

Solar energy is one of the most ubiquitous and abundant sources of energy on Earth. The Sun emits approximately 174 Petawatts (PW) of radiant energy every day, approximately 30% of which is reflected back into space, while the remainder hits the Earth's surface.

Although solar energy harvesting has been around for many years, it is only recently that we have begun to fully comprehend its potential. Approximately only 0.0005% of the Earth's energy consumption is met by solar power, signifying there is still a lot left to harness.

The basic framework for solar power harvesting involves the utilization of photovoltaic cells. The operating principle is straightforward - these cells convert sunlight directly into electricity by the photovoltaic effect. Solar panels composed of many cells are used to generate usable power quantities.

The effectiveness of solar energy harnessing relies on several factors. The intensity of sunlight, atmospheric conditions, angle of incident light, and the efficiency of the photovoltaic cells, all cumulatively contribute to the power output. Various emerging technologies, such as utilizing organic materials, multijunction cells, and solar concentrators, are being worked upon to augment the efficiency of solar energy harvesting.

Despite the assurance of inexhaustible energy, solar power does face certain challenges. These primarily pertain to the intermittent nature of sunlight and the substantial space required for erecting solar panels.

5.2. Wind Energy Harvesting

Wind energy, converted into mechanical energy for hundreds of years now, has been successfully harnessed for electricity production over the last century. Wind energy manifestation relies on the kinetic energy transferred from wind to the blades of a wind turbine.

Wind energy has emerged as one of the most rapidly expanding forms of renewable energy in recent decades. The improvement in wind turbine technology and reduction in costs has driven the escalation. Wind energy is profuse, renewable, widely distributed, and releases no greenhouse gases during operation.

However, similar to solar energy, wind energy is heavily influenced by environmental and meteorological conditions. The intermittent nature of wind patterns, coupled with the requirement of building these massive turbines in open areas often far from heavily populated regions, limits its widespread adoption.

Research is ongoing to devise novel wind energy harvesting technologies, such as airborne wind energy (AWE) systems, that can capture the immense potential of high-altitude winds.

5.3. Kinetic Energy Harvesting

Kinetic energy harvesting captures energy from motion and turns it into electricity. This can be from a variety of sources like human activity, vehicle motion, water flow, and even vibrations. The basic principle involved in kinetic energy harvesting is the application of piezoelectric or electromagnetism.

The opportunities for kinetic energy harvesting are numerous, primarily due to its pervasive nature – motion is an inalienable part of our everyday lives. Micro kinetic energy harvesting devices are effectively used in wearable techs as power sources. Larger applications include capturing energy from ocean waves or vehicular motion.

The challenge with kinetic energy harvesting lies in the irregularity of motion sources and the relatively small amount of energy captured at once. Advanced materials and nanotechnology hold the key to enhance the efficiency of kinetic energy harvesting.

5.4. Comparative Analysis

While solar and wind energies are region-specific, kinetic energy can be harvested almost anywhere. However, currently, solar and wind energy harvesting technologies provide larger power outputs compared to kinetic energy.

Solar energy yields a constant source of power on bright days, while wind energy is viable both day and night, given ideal climate conditions. Conversely, kinetic energy can be harnessed anytime, but the amount of energy generated is proportional to the intensity of the motion.

All three energy harvesting methods have their specific limitations and challenges, but with continuous innovations, they are all moving closer to becoming more plausible, sustainable, and versatile power sources.

In conclusion, solar, wind, and kinetic energy harvesting form the triptych of sustainable power generation. Each has unique advantages and challenges, but they collectively represent a way forward towards a cleaner, greener, and more sustainable future.

Chapter 6. Energy Harvesters: Environmental Impact and Sustainability

Global energy consumption is projected to increase in the subsequent decades, with the primary driving factors being population growth and economic expansion. The prevalent energy model, currently reliant on fossil fuels such as coal, oil, and gas, is perilously unsustainable. As energy demand rises, so too do the environmental costs associated with high-carbon fuels. Enter the realm of Energy Harvesters, a promising option for generating power that is eco-friendly and sustainable.

6.1. The Environmental Impact of Traditional Energy Sources

Conventional energy sources often have significant environmental implications. The extraction, processing, transport, and burning of fossil fuels lead to a release of large amounts of greenhouse gases (GHGs) into the atmosphere, prompting global warming. Coal plants, for example, emit high levels of sulfur dioxide, leading to acid rain and respiratory disease. Oil spills from transport and extraction operations can cause extensive damage to marine ecosystems. Natural gas, although cleaner than coal or oil, still contributes to carbon emissions and can result in methane leaks, a potent GHG.

Adverse effects extend beyond emissions. Extraction operations often cause substantial harm to local ecosystems. Mining destroys habitats and can lead to soil erosion and water contamination. The water used in energy production processes also strains the availability of this critical resource for other uses.

Energy Harvesters present a potential solution to these problems.

6.2. Energy Harvesters: An Overview

Energy Harvesters are devices designed to capture and convert ambient energy sources into useful electrical energy. These ambient energy sources can include solar or light energy, thermal energy, wind or fluid energy, and even kinetic or mechanical energy.

For instance, vibration energy harvesters can convert mechanical energy from machines, buildings, and human movements into electrical power. Thermoelectric energy harvesters can convert waste heat energy from industrial processes, vehicle exhausts, or natural thermal gradients into electricity. Piezoelectric materials can convert pressure changes from acoustic or ultrasonic waves, mechanical vibrations, or even bodily movements into electrical energy.

6.3. Energy Harvesters: A Sustainable Solution

The desirability of energy harvesters as a source of power lies in their sustainability. By utilising otherwise wasted or inaccessible energy, they offer a way to generate electricity with less environmental degradation. But to what extent can they alleviate the environmental impact of our power needs?

To answer this question, we'll discuss some of the critical aspects of the environmental impact and sustainability of energy harvesters.

6.4. Reducing Greenhouse Gas Emissions

By directly converting renewable and often wasted ambient energy into power, energy harvesters can reduce carbon emissions and help mitigate global warming. Unlike traditional energy generation, ambient energy harvesting processes do not involve combustion or any reaction that produces GHGs.

6.5. Mitigating Environmental Disruption

Unlike extraction operations for fossil fuels, which can disrupt ecosystems, energy harvesters can often be placed within existing environments with minimal impact. For instance, vibration energy harvesters can be incorporated into current structures or machines without changing their external configuration or function significantly.

6.6. Lowering Energy Waste

Energy waste is a crucial problem in our current energy system. Energy harvesters can offer a solution to this by turning wasted energy into useful power. For instance, thermoelectric harvesters can convert waste heat from industries into usable electricity, potentially increasing overall energy efficiency significantly.

6.7. Promoting Energy Independence

On a larger scale, the widespread use of energy harvesters could decrease our reliance on large centralized power plants and foreign oil, fostering energy independence. Harvesting ambient energy is possible in many settings and thus could enable decentralized, local power generation. This would decrease transport needs, further reducing carbon emissions and enhancing resilience against power disruptions.

6.8. Energy Harvesters: The Road Ahead

There are obvious environmental advantages to using Energy Harvesters. However, certain technical and economic challenges must be overcome before their widespread adoption. These include enhancing the efficiency of energy conversion, reducing the cost of energy harvesting devices, and integrating these devices into existing and future infrastructures.

Despite these challenges, the potential benefits of harnessing ambient energy sources are too great to dismiss. Energy Harvesters represent a new frontier in sustainable energy, presenting an innovative path for power generation that is both green and sustainable. They could play a vital role in our move towards a more sustainable energy future, where the environmental costs of power generation cease to be a limiting factor on our prosperity and well-being.

In conclusion, as we gather momentum in our transition to more sustainable energy models, Energy Harvesters offer a promising prospect. They capture the hitherto untapped energy around us and convert it into usable power, shrinking our carbon footprints and inflating our hopes for a sustainable tomorrow.

Chapter 7. Innovative Applications of Energy Harvesting in Today's World

With the increasing awareness about and commitment to sustainability, energy harvesting technologies have become the cornerstone of a new era in power generation. They siphon energy from the ambient environment around us —heat, light, motion, etc.— and convert it into usable electrical energy. The innovative applications of such technologies today range widely, from self-powered sensor networks to wearable electronics. The prosperity of our interconnected world heavily rests on the shoulders of these advancements.

7.1. Harvesting Human Energy

People generate energy through day-to-day actions like walking, breathing, heat produced from our body, and even while we sleep. This human-generated energy, largely going to waste now, can be harvested using piezoelectric and thermoelectric devices. These devices convert mechanical and heat pressures into electricity.

For instance, the gait-based energy harvesting systems embed piezoelectric materials within shoe soles. When force is exerted while walking or running, these materials produce electrical energy. Similarly, thermoelectric harvesters utilize body heat, enabling applications such as self-charging fitness trackers or medical monitors.

With the proliferation of wearable tech, industries stand on the brink of a massive shift towards self-powered electronics. The progress in the field of wearable, flexible, and stretchable energy harvesters is poised to replace conventional batteries in near future and has taken us a step closer to a truly wireless world.

7.2. From Building Facades to Energy Reservoirs

Another prominent application of energy harvesting is seen in modern infrastructure. Buildings–especially skyscrapers–cover large surface areas, offering a substantial canvas for energy harvesting. Photovoltaic (PV) glass, a key technology, transforms windows and glass facades into solar power generators. Innovative depictions of such technology can be found globally. For instance, Apple Inc.'s office in Cupertino, California, known as the "Apple Park", employs a 17-megawatt rooftop solar installation that it uses to run its facility 100% on renewable energy.

Even motion energy from elevators and revolving doors can be captured and converted. This energy self-sufficiency of buildings contributes to the development of smart cities—a central component of sustainable urban civilization.

7.3. Self-powered Sensor Networks

Across industries ranging from agriculture and healthcare to environmental monitoring and military, remote sensor networks are integral. Traditionally, these are powered using batteries, which demand periodic replacements or recharging. However, energy harvesting can offer an enduring solution.

Radio Frequency (RF) energy, vibration, and solar energy can be harvested to power these sensor networks. This way, the networks can function indefinitely without the need for human intervention. Data collection from remote regions becomes feasible and cost-efficient with this approach. Also, in uncordial environments, such as battlefields or inaccessible regions, lifespan and reliability of sensors increase significantly.

7.4. Energy Harvesting in Transportation

With electric vehicles (EVs) shaping the future of transportation, energy harvesting has an essential role to play. While regenerative braking systems (which convert kinetic energy during deceleration into electrical energy) have been in use, the real advancement is seen in the exploration of energy harvesting roads.

These roads harvest the pressure exerted by moving vehicles through piezo-electricity. A promising pilot project in Israel, "Energypav: Asphalt", demonstrates how energy from a vehicle's weight can be harvested to generate electricity. Besides, thermal and solar energy from roads can also be harnessed, triggering a revolution in sustainable transportation infrastructure.

7.5. Harvested Energy for Internet of Things (IoT)

The globally interconnected mesh of devices, or IoT, demands uninterrupted power supply for seamless function. With billions of devices connected today, implementing conventional power solutions is neither feasible nor environmentally friendly. Energy harvesting comes to rescue, offering a sustainable solution.

The energy requirements of an IoT device are typically low, meaning the ambient energy around a device, be it light or heat, can power it sufficiently. Through process optimization, the lifetime and reliability of IoT devices could increase considerably. Plus, maintenance costs would reduce, as energy harvester-powered devices wouldn't require frequent battery replacements or charging.

7.6. Future Hopes and Aspirations

With a view to our future, energy harvesting promises self-powered, energy-efficient, and autonomous systems capable of revolutionizing our daily life. In this transition, strategic planning, research, and innovation can help overcome barriers like efficiency and adoption, paving a path to a sustainable future.

While this might sound highly futuristic, many applications are well beyond conceptual stages, with countless others innovating at breakneck speed. From "smart" routes for joggers that generate electricity to wearable tech powered through everyday movement, energy harvesters are far more than just promising—they are here and they are now. Only through continued research and a willingness to embrace these unconventional technologies can we hope to truly harness their potential.

Chapter 8. Challenges and Solutions in Energy Harvesting Technology

The promising capabilities of energy harvesting technology are matched by a circuitous set of challenges that engineers, researchers, and designers encounter when implementing these systems.

8.1. Efficiency and Energy Conversion

The most dominant challenge in energy harvesting is achieving optimum efficiency. Most energy sources for energy harvesters are naturally fluctuating and inconsistent, making them hard to harness optimally due to limitations in conversion techniques. For example, wind and solar energy are highly dependent on environmental conditions that can be unpredictable and variable. Achieving an efficient and effective energy harvesting system means designing a system that can capitalize on these fluctuating power levels without undermining the capability of the system to generate and store power.

Leapfrogging this challenge entails developing novel technologies that can adapt to naturally fluctuating energy sources. Dynamic power management systems enable energy harvesters to adjust their working conditions to align with the underlying source's energy level. Such adaptive techniques can significantly increase the overall efficiency of energy harvesting.

8.2. Size and Power Density

Another potent challenge is the size and power density of these technologies. For wearable and Internet of Things (IoT) applications, energy harvesting systems should be small, light, and capable of generating sufficient power. In many cases, they must achieve these goals without accessing an external power source.

The solution to these challenges lies in material science and engineering. New materials with high-energy density can provide an edge in designing small yet powerful energy harvesters. Similarly, advancements in fabrication techniques also play a significant role in reducing the size of energy harvesters.

8.3. Energy Storage and Management

Storing and managing harvested energy is a critical issue. Most of the energy we can scavenge in the environment is low-grade, meaning it's meager in quantity. This reality calls for the need for sophisticated energy storage and management systems.

Super-capacitors and advanced ultra-thin batteries may be a way forward. These technologies need further refinement, but they show promise by enabling long-term storage of small quantities of energy that can be employed appropriately and efficiently when needed.

8.4. Cost and Commercial Feasibility

A significant part of energy harvesting's future rests on the commercial feasibility of implementing energy harvesting technologies. High upfront cost of materials and fabrication processes stands as considerable obstacles before mass-market adoption to the technology.

Progress in scaling up manufacturing processes, making use of cost-effective materials, and designing efficient energy management systems can bring down overall costs and boost the commercial prospects of energy harvesting applications.

8.5. Standards and Regulations

The lack of standards and guidelines regarding energy harvesting applications, particularly where wearable and IoT devices are involved, poses a significant challenge since it adds uncertainty for manufacturers and designers.

Regulatory bodies need to develop comprehensive standards that take into account the various energy sources, harvesting techniques, and application scenarios, all while ensuring safety for end-users.

Despite these hurdles, energy harvesting technology is rapidly advancing, driven by a world that is increasingly conscious of the need for sustainability and innovative energy solutions. The potential of this technology is enormous, and navigating its complexities may be the key to unlocking a new era in power generation and sustainability. The journey is demanding, but with each falling barrier, we come closer to the dawn of an energy-independent world.

In conclusion, capitalizing on the untapped potential of our environment for generating power is prudent. Though energy harvesting is not without its challenges, innovative solutions on the horizon promise a future steeped in sustainability, efficiency, and energy independence. Our understanding and applications of this technology are evolving just as rapidly as the energy-consuming demands of our modern world - a tantalizing possibility indeed for the years to come.

Chapter 9. Key Market Players and Their Contributions

Energy harvesting technology, as an emerging field, has drawn several key players geared towards innovative contribution. The spectacular blend of research institutions, start-ups, and major conglomerates have shaped – and continue to shape – the industry's landscape.

9.1. Alphabet Energy

Alphabet Energy stands out as leading player in the thermoelectric sector. The company harnesses the waste heat produced during industrial processes and repurpose it into usable electricity, slashing both energy cost and carbon footprint. With their PowerCard™ technology, Alphabet Energy optimizes the conversion of waste heat into electrical power. The company's novel technology is designed for a broad array of applications, including cars, trucks, remote power sources, and industrial waste heat recovery, effectively democratizing access to energy-harvesting.

9.2. Seiko Epson Corporation

Often associated with its popular print and digital products, Seiko Epson Corporation, a Japanese company, has also forayed into the energy harvesting sector. Apart from traditional business, they have ventured into wearable technology by developing a GPS sports watch powered by light and body heat, enabling for longer battery life and more efficient use of power. Their continued research and product development in such areas imply the extent to which energy harvesting could influence major industries.

9.3. Powercast Corporation

Powercast Corporation is a pioneer in providing advanced RF-based wireless power technology and solutions. Its Powerharvester® receivers are embedded in various devices, harvesting energy from ambient RF energy sources to recharge these devices in a sustainable way. This represents a cutting-edge application of energy harvesting methodologies. Powercast's Lifetime Power® technology aspires to provide power for the lifetime of devices, changing the way devices are used and charged.

9.4. EnOcean

EnOcean has been leading the charge in the application of energy harvesting technology to building automation. Their technology uses energy from light, temperature differences, and even kinetic energy to power sensors and switches that in turn control lighting and HVAC systems. This helps reduce the energy footprint of buildings significantly while also saving on costs. Leveraging on their Dolphin platform, they deliver batteryless wireless technology, thereby saving maintenance costs, improving reliability, and enhancing overall energy efficiency.

9.5. ExoSystems

A relative newcomer, ExoSystems has created an energy harvesting device that harnesses electrical power from the vibrations generated in busy environments such as factories, transportations, and even bustling city streets. ExoCube, their groundbreaking offering, captures vibrations and transforms them into electrical energy. This plug-and-play solution, suitable for both indoor and outdoor applications, demonstrates the enormous potential of energy harvesting in smart cities and industrial IoT.

9.6. Kinergizer

Kinergizer, a Dutch energy harvesting company, focuses on developing motion energy harvesting solutions for IoT sensors eliminating the need for battery replacement. Their energy-harvesting system harnesses energy from shaking, swinging, and other small movements, enabling businesses to reduce their operational overhead in IoT management and promoting the longevity of devices.

These major players each contribute uniquely to the promising field of energy harvesting. Alphabet Energy focuses on recycling industrial waste heat, Seiko Epson employs thermoelectric technology to power wearables, Powercast provides RF-based wireless power, EnOcean uses ambient conditions to optimize building automation, ExoSystems leverages vibrations from urban settings, and Kinergizer targets motion energy for IoT sensors.

In the quest for sustainable power, energy harvesting is increasingly proving crucial. As this technology continues to evolve, these key market players – and likely many more yet unnamed – will help shape the new energy landscape. The contributions of each entity will guide the path towards fully embracing energy harvesting technologies in our daily lives, bringing forth a revolution in sustainable power generation. Their joint efforts promise to drastically reduce dependence on traditional, and often environmentally harmful, energy sources. They inspire hope for a future where the energy we use derives from the energy around us, in the most efficient and sustainable way possible. Further developments in the sector will depend on both ingenuity and investment, making this an exciting space to watch and invest in for the future.

Chapter 10. Future Projections: Energy Harvesting in the next Decade

As we gear up to bid goodbye to the 2020s, speculation around what the energy harvesting scenery would look like over the next decade is rife. Drawing from technological advancements, market trends, and shifting landscapes, this chapter endeavors to offer in-depth projections for energy harvesting in the next 10 years.

10.1. Shifting Energy Paradigms

Over time, the world's stance on energy usage has significantly evolved to incorporate more sustainable, renewable sources. The narrative no longer revolves around harnessing voluminous oil or coal reserves but has shifted towards optimizing energy extraction from available non-exhaustible resources. As we chart our path into the future, we're likely to witness a further evolution of this paradigm, with energy harvesters playing vital roles.

The ensuing decade will likely focus on decentralized energy production. Instead of relying on a few primary sources, the focus will be on utilizing smaller, personal sources of energy, much of which can be harvested on-site. This outlook will allow for the release of lesser greenhouse gases into the atmosphere compared to traditional energy generation methods. More so, it will be in line with the emerging concept of the 'Internet of Things', where devices could share energy as they share data.

Amidst the shifting energy paradigm, the onus will be on energy harvesters to enable such decentralized systems. There will be an increasing impetus on developing these technologies to convert different forms of energy available in our direct environment into usable power.

10.2. Technological Evolution

Progress in the field of nanotechnology is considered a driving force behind the evolution of energy harvesters. With the industry taking strides towards harnessing the capabilities of nanoparticles, the 2030s could see superior, more efficient mechanisms that better utilize omnipresent energy forms around us.

Incorporating nanotechnology can help design energy harvesters on a molecular level, giving rise to devices possessing higher energy density storage and conversion capacities than their predecessors. This approach might significantly reduce energy waste and contribute to a greener future.

The emergence of smart materials like piezoelectric materials, thermoelectrics, and photoactive organic compounds will undoubtedly have substantial impacts on energy harvesting. These materials contain properties that enable them to directly transform physical stimuli into electricity, laying the groundwork for innovative, self-powered systems.

Additionally, progress in the domains of micro-electro-mechanical systems (MEMS) and nanoscale manufacturing processes could pave the way for smaller, cost-effective energy harvesters.

10.3. Market Forces at Play

Market forces will play an instrumental role in directing the progress of energy harvesters in the coming decade. Increased awareness about environmental sustainability, legislation favoring renewable energy solutions, and subsidies encouraging innovation and adoption in the renewable sector will stimulate growth.

Projections from market researchers anticipate that the global energy harvesting market will witness exponential growth in the upcoming decade. The drive towards miniaturization, coupled with the need for sustainable power solutions for wearable technologies and IoT devices, will be notable contributors to this growth.

Strong competition among players in the energy harvesting market will also fuel substantial R&D efforts, leading to more efficient, affordable, and user-friendly devices. Moreover, as the demand for sustainable technologies escalates, so too will the pursuit of breakthroughs in the sector.

10.4. Making it Mainstream

Effectively tackling the challenges of energy harvesting—cost, efficiency, and scalability—can make these solutions more mainstream and adaptable in myriad applications.

Increased emphasis on commercializing energy harvesters will push for the development of solutions catering to an extensive range of sectors—be it healthcare, defense, industrial automation, or consumer electronics. Devices capable of converting thermal energy from body heat or kinetic energy from body movement into electricity could revolutionize the wearable technology space.

Thus, the future of energy harvesters isn't simply isolated to grand-scale implementations like wind farms or solar parks, but it permeates the fabric of our everyday lives.

10.5. Conclusion

The coming decade promises to anchor a new era in the realm of energy harvesting. As advanced technologies like nanotechnology and smart materials continue to converge with market forces and shifting energy paradigms, the horizon is bright for energy harvesters. With an overwhelming focus on sustainability, efficiency, and decentralization, the energy landscape of the 2030s could be vastly different from anything we've ever imagined.

Chapter 11. Closing Remarks: The Ultimate Impact of Energy Harvesters on Sustainable Power

The pivot towards sustainability is a decidedly global effort, and the introduction of energy harvesters on this stage is an undeniable game-changer. With the capacity to convert ambient energy into usable power, energy harvesters bring a wave of change that holds potential to reshape power consumption patterns at an unprecedented scale. By the same token, however, they also bring a maze of new challenges, raising questions about the feasibility of their widespread adaptation and about the precise manner in which they might influence the course of sustainable power in the years to come.

11.1. Harnessing The Omnipresent Energy

From a technological standpoint, energy harvesters are at the forefront of innovation. Designed to tap into latent energy around us, these devices capture power from various sources, such as solar, thermal, wind, and kinetic energy. The capacity to harness these resources and convert them into electricity opens up possibilities for self-powered electronic systems. In turn, this could significantly reduce our dependence on conventional energy sources, presenting a deeply attractive proposition for a world invested in combating climate change.

But the effectiveness of energy harvesters is not limited to their innovative design. Context, too, plays a pivotal role in determining their impact, and it is here that they truly shine. Energy harvesting techniques eschew the need for power grid connectivity, enabling them to serve remote locations otherwise denied this vital resource. Equipped with energy harvesters, these regions have the opportunity to access sustainable power even in the absence of mainstream infrastructural support.

11.2. Transitioning Towards Energy Efficiency

As environomically attractive as energy harvesters may be, their potential extends beyond mere environmental impact. Economic considerations, too, are inextricable from this discussion, particularly when evaluated from a long-term perspective. Reducing dependence on fossil fuels, and the costs associated therewith, energy harvesters could herald a significant decrease in power related expenditure on a global scale.

Retrofitting existing machinery and infrastructures with energy harvesting mechanisms can improve their efficiency multi-fold. For example, utilizing energy harvesters in industrial machinery can offset energy consumption by recovering heat or vibration energy otherwise lost. Such adaptations to efficiencies could streamline operational costs and usher in a new era of industrial energy optimization.

11.3. Paving the Way for a Sustainable Future

That being said, the ultimate impact of energy harvesters is not confined to the economic or the environmental; their sociopolitical significance is equally profound. The decentralized energy production they facilitate is particularly notable, challenging the established paradigm of power generation and distribution by negating the need for an extensive centralized production system.

In practical terms, this could level the playing field, empowering local communities with the capacity to generate and control their own power supply. Energy democratization might help alleviate energy poverty and contribute to socio-economic stability and development, especially in underprivileged societies.

Equally important is their influence in the policy-making arena. By presenting a model of decentralized, sustainable energy consumption, energy harvesters might well shape future directions of our political landscapes. Environmentally conscious policymaking may change the behaviors of industrial giants, nations, and even individuals, leading to a more sustainable global society.

11.4. The Road Ahead

It is, however, essential to temper our enthusiasm with pragmatism. While the potential of energy harvesters is vast, their integration into the daily lives of people is still a work in progress. Technological, economical, and logistical challenges lie ahead.

For energy harvesters to effectuate the revolution they promise, ongoing investment in research and development is crucial. Emerging technologies must be refined, costs must be reduced, and logistics resolved. Only then can they become a norm rather than a niche in the world of sustainable power.

In conclusion, there can be little doubt that the ultimate impact of energy harvesters on sustainable power will be transformative. They hold the promise to reshape our relationships with energy, driving us towards a future that is not merely 'sustainable' in the broadest sense of the term, but also equitable and efficient. Despite the challenges and complexities that lie ahead, the journey towards a world powered by energy harvesters promises to be an exciting one. The dream may be still out of reach, but as every step taken towards realizing it brings us closer to an era of sustainable power, the pursuit is undoubtedly worthwhile.

39